MW01365858

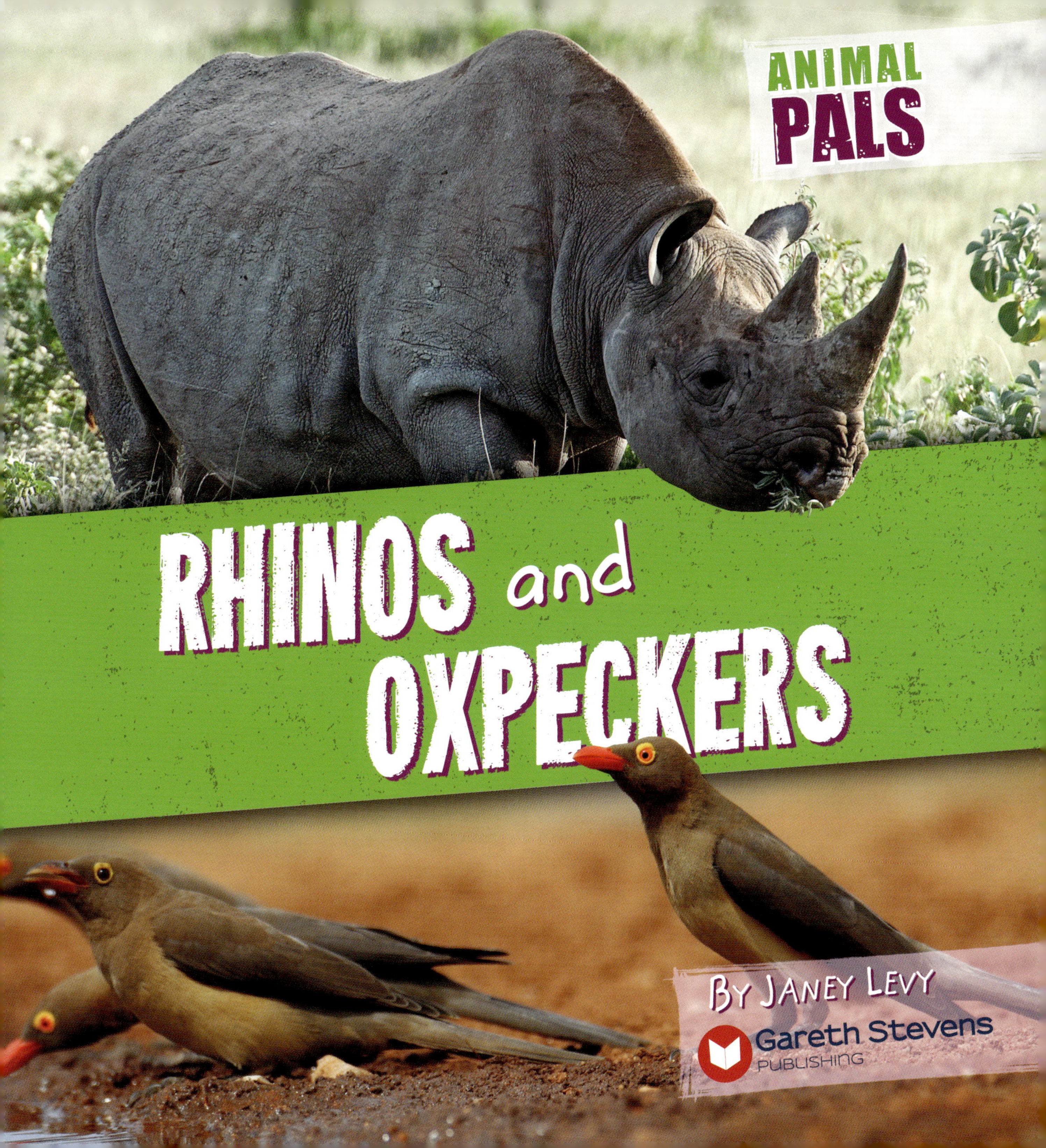
ANIMAL
PALS
RHINOS and
OXPECKERS
By Janey Levy
Gareth Stevens
PUBLISHING

Please visit our website, www.garethstevens.com. For a free color catalog of all our high-quality books, call toll free 1-800-542-2595 or fax 1-877-542-2596.

Library of Congress Cataloging-in-Publication Data

Names: Levy, Janey, author.
Title: Rhinos and oxpeckers / Janey Levy.
Description: New York : Gareth Stevens Publishing, [2022] | Series: Animal pals | Includes index.
Identifiers: LCCN 2020034907 (print) | LCCN 2020034908 (ebook) | ISBN 9781538266854 (library binding) | ISBN 9781538266830 (paperback) | ISBN 9781538266847 (set) | ISBN 9781538266861 (ebook)
Subjects: LCSH: Rhinoceroses–Juvenile literature.
Classification: LCC QL737.U63 L48 2022 (print) | LCC QL737.U63 (ebook) | DDC 599.66/8–dc23
LC record available at https://lccn.loc.gov/2020034907
LC ebook record available at https://lccn.loc.gov/2020034908

First Edition

Published in 2022 by
Gareth Stevens Publishing
29 E. 21st Street
New York, NY 10010

Designer: Andrea Davison-Bartolotta
Editor: Monika Davies

Photo credits: Cover, p. 1 (top) Sylvain Bonaldi/500px/Getty Images; cover, p. 1 Jurgens Potgieter/Shutterstock.com; p. 5 Martin Harvey/The Image Bank/Getty Images; p. 7 (main) pingebat/Shutterstock.com; p. 7 (inset) Handmade Pictures/Shutterstock.com; p. 9 (top) Arthur Morris/Corbis Documentary/Getty Images; p. 9 (bottom) Vicki Jauron, Babylon and Beyond Photography/Moment/Getty Images; p. 11 Rocco Umbescheidt/iStock/Getty Images Plus/Getty Images; p. 12 David OBrien/Shutterstock.com; p. 13 Simoneemanphotography/iStock/Getty Image Plus; p. 15 Roger de la Harpe/Gallo Images/Getty Images Plus/Getty Images; p. 16 matthieu Gallet/Shutterstock.com; p. 17 hpimagelibrary/Gallo Images/Getty Images Plus/Getty Images; p. 19 ZambeziShark/iStock/Getty Images Plus/Getty Images; p. 21 (top) Sylvain Cordier/Gamma-Rapho/Getty Images; p. 21 (bottom) slowmotiongli/Shutterstock.com.

Printed in the United States of America

CPSIA compliance information: Batch #CSGS22: For further information contact Gareth Stevens, New York, New York at 1-800-542-2595.

CONTENTS

Words in the glossary appear in **bold** type the first time they are used in the text.

SAVANNA SIDEKICKS

Rhinos are a common sight at zoos. But the wild home for many of these huge herbivores, or plant eaters, is the African **savanna**. If you've ever seen a picture of a wild rhino, you might have noticed a small bird on its back. What's that about?

That bird may be an oxpecker. It's long been believed rhinos and oxpeckers have a **relationship** that benefits both of them. But recent studies suggest their relationship isn't so simple. You'll learn lots more about rhinos and oxpeckers inside this book.

FACT FINDER!

A mutualistic relationship is a relationship between two different kinds of animals that benefits both of them.

How does an oxpecker stay on a rhino's back? It has very sharp claws that it uses to hold on to the rhino's thick skin.

REGARDING RHINOS

If you've seen rhinos at a zoo, you know these **mammals** are huge! Five **species** of rhinos exist. Africa is home to two species, including the white rhino. The white rhino is the largest species, measuring up to 13 feet (4 m) long and 6 feet (1.8 m) tall. A white rhino can weigh up to about 6,000 pounds (2,720 kg)!

Rhinos' horns are their most famous feature. Their horns are made from matter called keratin. That same matter is found in your hair and nails!

FACT FINDER!

Rhinos have excellent senses of hearing and smell but poor eyesight. Since they can't see well, they may charge at anything that surprises them—and that can be scary!

Where Africa's Rhinos Live

Africa

White rhinos and black rhinos are the two kinds of African rhinos. This map shows where they live.

white rhino

ABOUT OXPECKERS

While rhinos are huge, oxpeckers are tiny. Oxpeckers are a bit smaller than American robins. They measure only about 8 inches (20 cm) long.

There are two kinds of oxpeckers: red-billed and yellow-billed. They're plain birds with brownish upper parts, a tan or pale yellow stomach, and gray legs. An oxpecker's most outstanding features is its small, thick bill and red eyes with yellow circles around them. The red-billed oxpecker has—of course—a red bill. The yellow-billed oxpecker's bill is yellow with a red point.

FACT FINDER!

Oxpeckers are also known as tickbirds. That's because they eat ticks off the skin of rhinos and other animals. Yuck!

Oxpeckers have stiff tail feathers that help them balance as they ride along on the backs of rhinos or other animals.

RELIEF AND WARNING

If the relationship between rhinos and oxpeckers is mutualistic, then what do rhinos get out of it? Oxpeckers clean ticks and other **parasites** off rhinos. They remove dead skin and also eat earwax, which might help rhinos hear better.

Oxpeckers also warn rhinos of approaching danger. With their poor eyesight, rhinos often aren't aware of nearby enemies, such as human hunters. But oxpeckers spot these dangers more easily. Their sharp call serves as an alarm and gives rhinos time to escape.

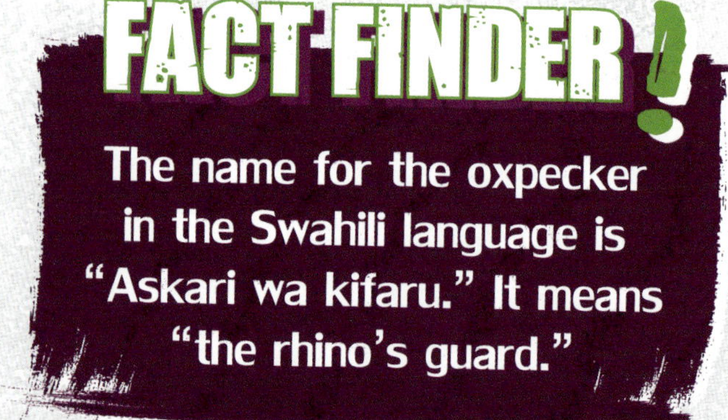

FACT FINDER!
The name for the oxpecker in the Swahili language is "Askari wa kifaru." It means "the rhino's guard."

Research shows that rhinos with oxpeckers often notice nearby people, while rhinos on their own often miss spotting humans.

A PICNIC

What do oxpeckers get out of their relationship with rhinos? For the birds, this is a necessary relationship. Oxpeckers can't survive without the food they get from riding on the backs of rhinos and other African animals.

When they ride on another animal's back, oxpeckers enjoy a feast filled with tasty foods. Oxpeckers have meals of ticks, other parasites, dead skin, and earwax. They also eat flies, **mucus**, and blood from open wounds. Yuck!

red-billed oxpecker drinking rhino drool

Would you enjoy eating like an oxpecker?

FACT FINDER!

Some oxpeckers pull hair from rhinos' ears to use to build their nests. Ouch!

ABOUT THOSE TICKS

Ticks suck blood from their animal **hosts** and can pass on **diseases**. It sounds helpful that oxpeckers eat ticks on rhinos, doesn't it? But it turns out there's more to the story.

Oxpeckers usually only eat one tick species and avoid other ticks on the rhinos. Most of the time, they eat the ticks once the ticks are full of blood from the rhino. That's because oxpeckers prefer blood as food. This may mean oxpeckers aren't truly saving rhinos from the harm caused by ticks.

FACT FINDER!

To discover if oxpeckers reduced the number of ticks on animals, scientists studied two groups of cattle. One group was paired with oxpeckers and the other was not. Both groups wound up with about the same number of ticks!

When choosing their meals, oxpeckers like to eat female blue ticks.

OTHER OXPECKER ACTS

Oxpeckers like to eat ticks full of blood. But they also drink blood directly from open wounds on rhinos. If oxpeckers find rhino wounds that are beginning to heal, they'll peck at them to make the wounds bleed again. They then have fresh blood to drink. This means it takes the rhino's wounds longer to heal completely.

Drinking blood from existing wounds is bad enough. But oxpeckers will sometimes make fresh wounds on rhinos to get blood.

Here, oxpeckers are gathered near a wound on a rhino's back.

FRIENDS OR PARASITES?

Scientists long believed rhinos and oxpeckers had a mutualistic relationship. That means oxpeckers and rhinos have a relationship a bit like friends. And friends help each other out. Oxpeckers do help rhinos in many ways. They clear off dead skin and eat pests off rhinos. The birds also warn rhinos of human **threats**.

But oxpeckers also act like parasites. They cause rhinos' wounds to bleed further and sometimes even create new wounds. So, do rhinos see oxpeckers as friends, parasites—or both? What do you think?

Is this a picture of a happy friendship? Or is one creature taking advantage of the other?

MORE BIRD-MAMMAL RELATIONSHIPS

You might think the relationship between rhinos and oxpeckers is unusual. But many other relationships between birds and mammals exist in the world.

Birds called cattle egrets are found across the world. They eat ticks off many large mammals, including different kinds of livestock and elephants. In the South American country of Brazil, birds called wattled jacanas and shiny cowbirds eat pests off giant **rodents** called capybaras. These mutualistic relationships exist because they work well for both creatures!

FACT FINDER!

In Turkey, scientists discovered frogs riding on top of water buffaloes. These frogs eat bugs living on the buffaloes!

The animal kingdom is full of birds and mammals working together to survive.

GLOSSARY

disease: illness

host: the animal on or in which a parasite lives

mammal: a warm-blooded animal that has a backbone and hair, breathes air, and feeds milk to its young

mucus: a thick slime produced by the bodies of many animals

parasite: a living thing that lives in, on, or with another living thing and often harms it

relationship: a connection between two living things

research: study to find something new

rodent: a small, furry animal with large front teeth, such as a mouse or rat

savanna: a grassland with scattered patches of trees

species: a group of plants or animals that are all of the same kind

threat: something likely to cause harm

FOR MORE INFORMATION

Books

Kirk, Daniel. *Rhino in the House: The True Story of Saving Samia.* New York, NY: Abrams Books for Young Readers, 2017.

Meeker, Clare Hodgson. *Rhino Rescue!: And More True Stories of Saving Animals.* Washington, D.C.: National Geographic, 2016.

Schuetz, Kari. *Cape Buffalo and Oxpeckers.* Minneapolis, MN: Bellwether Media, 2019.

Websites

Rhinoceros
kids.sandiegozoo.org/animals/rhinoceros
Find out more fun facts about rhinos at the San Diego Zoo's website.

Rhino Facts!
www.natgeokids.com/za/discover/animals/general-animals/rhinoceros-facts/
Discover more about rhinos at this website.

Symbiosis
kids.britannica.com/kids/article/symbiosis/400286
Learn more about mutualistic and parasitic relationships between creatures here.

Publisher's note to educators and parents: Our editors have carefully reviewed these websites to ensure that they are suitable for students. Many websites change frequently, however, and we cannot guarantee that a site's future contents will continue to meet our high standards of quality and educational value. Be advised that students should be closely supervised whenever they access the internet.

INDEX